Caring for Your Pet Dinosaur

Taking Care of Your TRICERATOPS

Gail Terp

BLACK RABBIT BOOKS

Hi Jinx is published by Black Rabbit Books
P.O. Box 227, Mankato, Minnesota, 56002.
www.blackrabbitbooks.com
Copyright © 2023 Black Rabbit Books

Gina Kammer & Marysa Storm editors;
Michael Sellner, designer and photo researcher

Library of Congress Cataloging-in-Publication Data
Names: Terp, Gail, 1951- author.
Title: Taking care of your triceratops / by Gail Terp.
Description: Mankato, Minnesota :
Black Rabbit Books, [2023] |
Series: Hi jinx. Caring for your pet dinosaur |
Includes bibliographical references
and index. | Audience: Ages 8-12 |
Audience: Grades 4-6 | Summary: "Struggling
or reluctant readers will laugh and learn as
they explore what it might be like to have a
pet Triceratops. From dealing with the
dinosaur's incredible size to figuring out how
to feed it, owning a dino is hard work!"–
Provided by publisher.
Identifiers: LCCN 2020042168 (print) |
LCCN 2020042169 (ebook) | ISBN 9781623106966
(hardcover) | ISBN 9781644665572 (paperback) |
ISBN 9781623107024 (ebook)
Subjects: LCSH: Triceratops–Juvenile literature.
Classification: LCC QE862.O65 T477 2022 (print) |
LCC QE862.O65 (ebook) | DDC 567.915/8-dc23
LC record available at https://lccn.loc.gov/2020042168
LC ebook record available at https://lccn.loc.gov/2020042169

Image Credits

Dreamstime: Ilona Orkin, 4; Kenneth Benner, 1,
19; Suryadi Djasman Kartodiwiryo, 12; Tigatelu, 11;
Shutterstock: 75ChuanStudios, 19; alexmstudio, 14,
15; Aluna1, 12; Angeliki Vel, 1; azmeyart, 8; Christos
Georghiou, 16; Denis Cristo, 11; ekler, 10; Haso,
11; IreneArt, Cover, 3, 8, 21; mejnak, 4, 11; Memo
Angeles, Cover, 1, 4, 5, 6, 8, 12, 14, 15, 16, 17, 19;
nikiteev_konstantin, 17; Pasko Maksim, 16, 23, 24;
picoStudio, 4; Pitju, 15, 21; Ron Dale, 5, 6, 13, 20;
Sujono sujono, 7, 12, 23; Teguh Mujiono, 3, 6, 7, 20,
21, 23

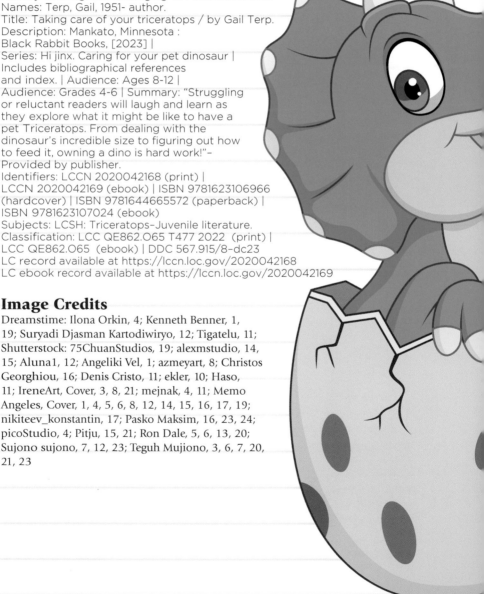

CONTENTS

Chapter 1

Is a TRICERATOPS Right for You?

With huge horns and a nice **attitude**, a Triceratops makes a great pet. You can ride it around your yard. Your family can play ring toss games using its horns. Owning a Triceratops isn't without **risks**, though. These large animals can cause a lot of trouble. If they start swinging their 8-foot- (2-meter-) long heads, watch out! They also need a lot of care. Read on to find out if one is right for you.

Chapter 2
Understanding Your TRICERATOPS

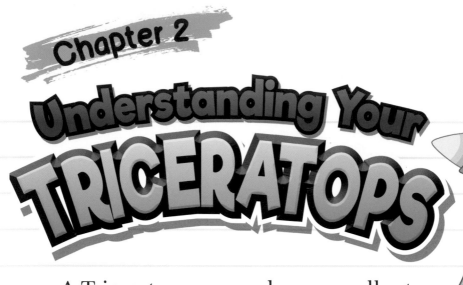

A Triceratops can make an excellent pet because of its fun personality. These dinosaurs enjoy all sorts of games. Ring toss is a favorite, but hide-and-seek is good too. A Triceratops loves playing dress-up as well. You'll need to play with your Triceratops often to keep it happy.

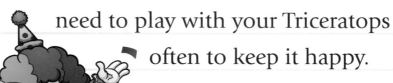

T. rex

barking dogs

rustling leaves

8

Sometimes Fearful

You should also know that lots of things scare a Triceratops. Strong winds rustling leaves. Neighborhood dogs barking. They can all frighten a Triceratops. And a scared Triceratops is a dangerous thing. The dino can run around in circles, shaking the ground and rattling buildings. You'll need to be understanding and patient. Speak soft, **soothing** words to your pet. And never mention T. rexes. A T. rex likes nothing better than eating a Triceratops.

Scientists have found fossilized T. rex poop with Triceratops bones in it.

Friendly Creatures

Scientists think these dinos lived in small family groups in the wild. Yours will want some company. Spend time talking with it. Put a picnic table near its **trough** and have meals with it. If you have enough room, you might want to get a second Triceratops. That's twice the work, though!

Triceratops is the state dinosaur of Wyoming. The dino is proud of this fact. Be sure to bring it up often.

Chapter 3

Caring for Your TRICERATOPS

A Triceratops needs a lot of care. It's a big dinosaur, and it doesn't like being in tight spaces. It doesn't need a big barn, but a huge yard is important. Don't have a huge yard? Make sure you live near a field or dog park. These dinos need plenty of room to run. They can reach 10 miles (16 kilometers) per hour.

A young Triceratops might seem manageable. But fully grown, it'll be about 12,000 pounds (5,443 kilograms) and 30 feet (9 m) long.

Grooming

Triceratops' large body needs grooming. To start, its teeth and horns will need brushing. Use a large brush with a long handle. A bath brush works perfectly. It's also great for washing your pet's back. There's a lot of skin to clean!

A Triceratops can have up to 800 teeth in its lifetime. Give yourself plenty of time for teeth brushing.

The front of a Triceratops' mouth
is a hard beak. It uses this beak
to grab and pull up plants.

Feeding

Big animals need big meals. Your Triceratops will spend most of the day munching on plants. But that can get expensive. Do you live near a farm? See if local farmers will give you the weeds from their fields. Or you could plant your own garden.

Amazing Friends

Triceratops are great fun as pets. But there's a lot of things to consider. Do you live near a big park? Can you help your Triceratops feel safe? Will you be able to feed it enough? If you answered yes, great! Triceratops is perfect for you.

Chapter 4

Get in on the HI JINX

Triceratops has gone **extinct**. No one can keep one as a pet now. But **paleontologists** dig up and study Triceratops bones. One dig site is in Montana. There, researchers discovered more than 50 Triceratops skulls. They study these skulls, learning more about these creatures. Maybe someday you'll study Triceratops fossils.

Take It One Step More

1. Would you like to study dinosaur bones when you grow up? Why or why not?

2. Plan a meal to share with your pet Triceratops. What would the two of you eat?

3. What do you think would be the hardest part about owning a Triceratops?

INDEX